John S. Billings, Theodore Rue

Diseases of the Respiratory Organs

Being original ideas advanced by a close student of nature and her

mysteries

John S. Billings, Theodore Rue

Diseases of the Respiratory Organs
Being original ideas advanced by a close student of nature and her mysteries

ISBN/EAN: 9783337198268

Printed in Europe, USA, Canada, Australia, Japan

Cover: Foto ©berggeist007 / pixelio.de

More available books at **www.hansebooks.com**

DISEASES

Beverly N. J.
Feby 23/92

Dr John S Billings
          Dear Sir.
                    I take
the liberty of sending you
a copy of a book lately
written by me which you
might find time to peruse
and if only one Idea in
it is new it might be of
Value to you in your new
duties — Your name is very
familiar to me — as nearly
all your Vouchers paid on
a/c of the army passed thro
my hands as auditor
          I was with Genl Kuckler

Crossman, Boyd, Brown &c in
this City & 2 yrs in the
Treasury Dept —

I only published a
few Copies of this work — but
intend to get a larger
Edition of new matter
& revise. or put in back
of this — I will have many
New Ideas in my next book
& will be pleased to
send a Copy to you —
                    — Very Respectfully
                         Theodore Kee

PREFACE.

It is possible that the medical profession may regard the publication of this little book as an unwarrantable intrusion upon their own territory. When, however, they consider how widely and for how many ages doctors have differed as to the causes of the complaints herein discussed, they may be willing to excuse a man from the people for stepping in to settle the dispute. The intimate relation which these diseases bear to our mode of life and the sanitary condition of our dwellings has long been known. This relation, universally acknowledged, but imperfectly understood, has caused the writer deep thought, the result of which is the present treatise.

A physician's only end and aim is too often to effect a cure, not to ascertain first the exact cause of the illness. But people should first be instructed in the causes of disease; they will then know how to avoid them. The writer knows nothing of medicine, excepting the remedies that are Nature's own, and has, consequently, avoided all medical terms and technical phrases; and while no one could appreciate more highly than he the attainments and services of the members of the medical profession, yet he ventures the opinion that Nature may reveal some of her secrets of health and disease to an earnest student outside the halls of a medical college perhaps more readily than to one within. It will remain for physicians and scientists, who have the requisite knowledge and the appliances necessary for further investigation to prove the truth of what is here advanced, and to devise cures for the diseases discussed, their causes having been plainly stated.

What the writer wishes to bring most prominently to the reader's attention in the following pages is the unknown or unheeded fact that *a large proportion of the fine particles of white dust floating in the air of our dwellings are spores and fragments of microscopic fungi, which commonly grow on textile fabrics, and are inhaled and carried to every part of the respiratory system; if a clogging of the air-tubes follows, there is a consequent partial confinement of unwholesome air which promotes the germination and growth of the inhaled spores, unless they are expelled by exercise and vigorous respiration in the open air. A moist and malarial atmosphere increases this fungous growth, which is always in the system of the aged and those having any disease of the respiratory organs, while dry and pure air diminishes it.*

The appearance of this work was promised for a certain date, and the author was consequently limited in time in its preparation. He trusts that this fact will lead his readers to treat leniently any traces of inaccuracy or haste.

Consumption, Hay-Fever,

AND OTHER DISEASES OF THE

Respiratory System.

Every effect is traceable to a cause, and nature, if rightly interrogated, gives the causes of more diseases than all the books that were ever written. While doctors have been going to secondary sources for the causes of disease, I have had recourse to nature directly, delving in nooks and by-ways, of which the physicians were ignorant, or from which they would be unable to derive any useful information. It is only from a wide and varied experience of life—to which I may lay some claim—that we learn to contemplate its phenomena intelligently. No one brought up in ease and affluence, as is the case with many of our physicians, can learn well the lessons that nature is ready to teach those who earnestly and persistently come to her for instruction. No mere learning of a profession or poring over books can produce those habits of keen observation which are indispensable to him who would make an important discovery.

Small and apparently trivial incidents have led to the greatest inventions and discoveries. Vaccination was first conceived of when it was noticed how the

5

sores on the teats of a milch cow were communicated to the hands of the milker, and from his hands, again, to the other cows in the herd. The principle of the electric telegraph was discovered by means of the simplest appliances—a kite and a key in the hands of Franklin. Watt conceived the steam-engine when he saw the lid of the tea-kettle agitated by the pressure of steam beneath it.

Before three generations shall have passed away, consumption and kindred affections of the throat and lungs will, I venture to prophesy, be spoken of as diseases of the past. This result will be effected through the intelligence of the people and by the passage of wise sanitary laws, when once the true cause of such maladies has been made known. Premature old age, which can be traced to the same cause, will be unknown. With one of the greatest enemies of human life and human health held at bay, the average age attained by mankind will be greatly increased, and a man will be considered only in his prime at the age of one hundred. Statistics show that in all civilized countries the ratio of births to the whole population is yearly decreasing. But even so, there need be no fear that the world will be depopulated. As the number of births decreases, the age of man will increase, through the eradication of those diseases of the respiratory organs which every year cause fifty per cent. of the deaths of the civilized world.

Every child born of healthy parents should die of old age. All other causes of physical decay are preventable in their nature, and can be avoided when they are once understood.

It was one of my tasks, when a boy, to keep the vegetable garden connected with our farm free from weeds. A more tiresome task, or one more productive of aching back and stiffness of joints, can scarcely be imagined. A discouraging, thankless piece of work it was, too; for no sooner was one end of the garden reached than the weeds would be seen springing up with a new lease of life at the other. Day after day and week after week I saw the unwelcome phenomenon repeated and the constant revival of the enemy I thought I had slain.

As a mere boy I learned well the lesson, how rapidly and persistently all harmful and parasitic forms of vegetable life will grow and multiply.

A thousand times more wonderful than the rapid growth and increase of weeds in the garden is the vigor and productiveness of that large class of plants —many of them microscopic in size—known by the name of *fungi*, and including all kinds of mould, mildew and smut. The most powerful microscope is needed in many cases to distinguish this fungous growth from the fabric or material on which it fastens itself; and even with the aid of the microscope, the whole secret of its origin and growth has not been discovered.

From Nees von Esenbeck in 1583 to Elias Frees in 1821, mycologists were puzzled to account for the seemingly spontaneous appearance of mould and mildew wherever there were given certain conditions of moisture and temperature. Many held the theory of spontaneous generation, not conceiving it possible that seeds or spores could be dispersed so widely or be

7

of such marvelous minuteness as observation and experience proved must be the case.

Researches of recent years have disproved both the theory of spontaneous growth and that of generation from chemical action. They have also shown that the varieties of microscopic fungi are practically infinite, and that their modes of reproduction vary from the simple process of division to the more complex apparatus of bud, blossom and seed-vessel. In the lower and more common varieties growth takes place by the addition of new cells to the parent germ, and each of these cells may, in turn, become the beginning of a new plant, on being detached and placed where there is sufficient warm, unwholesome air and moisture. Of the higher orders of fungi, the well known puff-ball, with its pod of millions of microscopic spores or seeds, furnishes a good example.

The rapidity of growth of all varieties of fungi, from the microscopic mildew to the largest mushroom, is almost inconceivable. Bits of paper-pulp thrown out scalding hot from the vat have been found covered and permeated with mould within twenty-four hours. A bar of iron, taken white-hot from the forge at night, has been discovered coated with mildew in the morning.

The speedy destruction which this growth works on both vegetable and animal tissues is to be noted. The mould which infests grain no sooner appears on the outside of the husk than it destroys the protective covering, by means which are probably both mechanical and chemical, and eats into the kernel itself.

This plant, which we call mould or mildew, has great power, small as it is, to do harm, as have some equally small species of animal life. If an elephant had the activity of a flea, in proportion to its size, it could leap across the ocean; so an insect or a plant, though many times smaller than the flea, may have a power for doing harm many times greater.

Wherever iron or steel will rust or silver tarnish, mould and mildew will grow; and these conditions are found in every dwelling, with its carpeted floors, its upholstered furniture and tapestried or papered walls.

Store-rooms and closets are packed with clothing and fabrics of every kind, and during the warmer months of the year no fire is kept excepting that needed for cooking in the kitchen, and this, of course, sends its heat through only a very small portion of the house. Accordingly the dampness entering the rooms on every rainy or foggy day is allowed to permeate every thread of carpet and stored clothing and to remain there, offering an ideal abode to mould and mildew until fires are started in the fall.

No conditions could be more favorable for the growth of mould than those existing in nearly every house during the summer. Carpets in particular, nailed to the floor as they are, and remaining so long unmoved, offer the best lurking-place for microscopic fungous growths. Between the carpet and the floor, and attaching itself to the fibres of the former, grows the mould which is so well known to housekeepers and which shows itself so clearly, when a carpet is taken up, as a fine dark powder on its lower side and on the floor.

Fully fifty per cent. of the premature deaths occurring in the civilized world are to be laid to the charge of carpets nailed to the floor, stuffed and upholstered furniture, certain styles of skin-tight and unventilated clothing, and the custom of taking wearing apparel, blankets, quilts, sheets and other fabrics out of closets and store-rooms, on the change of seasons, without properly drying, brushing, shaking, fumigating or otherwise cleaning preparatory to use.

No such safeguards are adopted, the damp, mildewed blankets and quilts being thrown at once on the bed and the musty, mould-infected garments hastily put on after a few strokes of the clothes-brush. No such thing as a thorough airing, shaking and drying, either in the sun or by a hot fire, is thought of. Consequently the mould particles and spores are left to be detached with every movement and set flying in the air, there to be inhaled by the occupants of the house. And this goes on for weeks or even months before even the thickest of the mould is beaten out and the garments or blankets thoroughly dried.

It is only during the last few centuries that these objectionable modes of house-furnishing and these reprehensible domestic practices have prevailed, and it is during this period that bronchial and pulmonary diseases have become so emphatically a scourge to our race.

That carpets are in reality such deadly haunts of disease and death as has just been asserted, will, we trust, be clearly demonstrated in the course of the present article.

10

Carpets, curtains, clothing, material used in upholstery, and other kinds of fabric are generally made with variegated hues for the express purpose of deceiving the eye as to their soiled condition. In fact, housekeepers often value a carpet because it "will not show the dirt." If, however, we were forced to use perfectly white or black fabrics in our house-furnishing, every good housekeeper would have her carpets, cushions and other such articles thoroughly cleaned every day or two. The fact that the pattern of a carpet conceals the dirt does not change its condition.

Matter of every kind, such as vermin and vegetable decay, finds its way to the carpet, being blown in at the windows, dropped on the floor, carried on the shoes, and otherwise deposited there, trodden down and ground up, until the fibres are filled with decaying and unwholesome organic and inorganic matter, and offer facilities for the maintenance of all sorts of fungous and noisome forms of life. Every drop of liquid spilt on the floor, as well as all dampness in the air, waters this miniature garden, and every muddy boot adds richness to its soil, while the housewife's broom only brushes the upper surface, and leaves the mass of the carpet and its under surface to support as many millions of microscopic fungi as can gain a foothold.

As was indicated above, the growth of this class of vegetation is inconceivably rapid. Millions and millions of spores and dead particles of mould are becoming detached every day, and are driven into the air in increasing numbers with every footfall and

11

every stroke of the broom, to be inhaled by the inmates of the house.

The longer the carpet remains, the greater the growth of mould. Often, when a carpet is removed, the floor is found to be covered to the depth of an eighth of an inch or more with fine black powder. This is, in part, dust from the street, but a large proportion of it is mould, being composed of the spores or seeds, filaments and roots of the little plants, and also dead vermin and animal matter, ground to powder by the treading of many feet.

On sending a ray of sunlight through a darkened room, and then causing some one to walk or run over the carpet, clouds of dust are seen to rise in the path of the sunbeam; or a dark lantern may be set on the floor, and the carpet lightly beaten, when the atmosphere in the light's path will appear thick with fine, floating particles. What is revealed by the ray of sunlight or lantern-light, is only an illustration of the daily condition of the atmosphere of a carpeted and occupied room.

The wonder is that we breathe so easily and unconsciously as we do, and that we are not more often stricken down with consumption, diphtheria, typhoid fever or other fatal diseases that gain access through the organs of respiration.

All fabrics, and not carpets alone, are favorite dwelling-places of mould and mildew. Examine the seams on the inside of a garment that has been worn only a month, and on those portions which touch the body and receive moisture from it will be found a growth of mould that will be taken for dust; but that it is not dust is proved by its absence on those parts of the garment that hang loose and receive no moisture from the body.

When we sit on a cushioned seat any length of time, sufficient moisture is imparted to the cushion to maintain a coating of mould as thin as the layer of vapor left on glass by our breath. On dying, the mould leaves its spores and dried fibres to be driven into the air, a source of danger and disease. Beat an upholstered chair, and you will see them rise in volumes. Some dust is intermixed, it is true, but the greater part is mould and its spores. The danger from fabric-mould is probably increased by the use, in nearly all kinds of fabrics, of poisonous aniline dyes which contribute to the nutriment of the mould.

If, by way of experiment, we lie on the floor and let several persons walk or run over the carpet, we are nearly smothered by the volumes of minute particles which start up at every foot-fall. We are affected less if we stand to our full height.

Children, however, are so short that they are obliged to suffer severely from these clouds of carpet mould; especially is this the case when they are so small as to be much of the time creeping or lying on the floor, often breathing directly through the mouth, so that the air has no chance to deposit a part of its poison before entering the throat and lungs. The child's organs of respiration, too, are not so strong and able to resist infection as are those of an adult.

No carpeting or upholstery should be allowed in any dwelling, excepting such as can be removed to the open air and thoroughly beaten and cleaned at least as often as once in every three days. It is well known what constant care is needed to keep a garden free from weeds. A lawn requires equal vigilance on the part of its keeper; he must go over it often, and pull up a weed here and there. To a casual observer the soil of a nicely kept lawn seems free from the slightest suspicion of weeds; but let it be turned up by the plough and left untouched during the growing season, and in an incredibly short time many kinds of weed will be seen growing in rank luxuriance. So the housekeeper may sweep and dust and remove every trace of mould and mildew from the outside, while beneath the carpet and in the stuffed chairs and lounges countless varieties of fungi are growing and depositing their spores.

If the house be neglected for a week the fungous growth will show itself to the eye in various kinds of mould and mildew, and if a month or several months have elapsed since the house was last occupied, as is the case with many dwellings and many rooms in the summer, a thorough cleaning and airing would be necessary to render it habitable for persons at all inclined to diseases of the throat or lungs.

All carpets and fabrics used in house-furnishing, as well as all stored clothing, should be beaten and cleaned, after a period of neglect, and fires should be made, even in hot weather, to dry thoroughly every room in the house which has been unoccupied for any length of time, before it is fit to be again lived in. Mattresses and pillows, too, especially after disuse and after they have become damp from the perspiration of

14

the human body, should be long and thoroughly aired or even taken to pieces and re-made.

Wall paper often becomes covered with a blue mould, where there is dampness and poor ventilation. The flour paste used by paper-hangers sours and moulds when it is first put on, and the spores of mould are thus left on the walls after they have become perfectly dry. As soon as the conditions of heat and moisture are again suitable, the mould springs into life once more. So newspapers and wrapping-paper quickly mildew when dampened, and after they become dry the spores are scattered in the air.

Churches, theatres and public halls are favorite lurking places for all kinds of mould, as they remain unventilated and darkened much of the time. Churches, in particular, being carpeted and cushioned and only opened once a week, are sure to become infected with every kind of unwholesome fungous growth.

Many of the parishioners in the country go to church wearing mouldy clothing, which they put on only once a week, driving in mouldy carriages, which see the sun only once in seven days, and sit on mouldy pew-cushions which are never exposed to the sun and air. After the commotion attending the gathering and seating of the congregation, and the consequent dislodging of mould particles, what wonder if the atmosphere becomes filled with poisonous germs and the minister catches the clergyman's sore throat after an hour or more of speaking!

Just as the plants and weeds of the fields grow most abundantly and rapidly in the months from May to September, so do all fungi at this time have the

15

greatest vigor of growth and reproduction. It is at this season that fires are discontinued in dwellings, while the dampness of night air, of fogs, and of showers is allowed to enter through open doors and windows. Rainy or damp weather may prevail for weeks at a time, and this condition of atmospheric moisture is especially marked near large rivers, lakes, or the ocean. The atmosphere, saturated from these sources, penetrates inland for miles. At a distance from these large bodies of water, consumption, hay-fever, and other diseases of the throat and lungs, are less common. The explanation is not that a moist atmosphere is in itself necessarily fatal to health, but that every variety of unwholesome fungous growth is fostered by this excess of humidity.

If dwellings were so furnished as to offer no attractive lurking-places for mould and mildew, the dampness would not endanger health, but would tend rather to equalize and modify the temperature, both in winter and summer.

As a young man, in the sixties, I was called, by the nature of my business, to every part of the country, spending sometimes three or four weeks in one place. On these occasions I secured board and lodging at some private house, instead of going to a hotel. It frequently happened that I needed certain papers from my room in the middle of the day, and on going for them I was pretty sure to find my chamber in perfect order, but shutters closed and curtains down, so that I had to light the gas to find what was required.

This illustrates the general custom of our housewives—to shut out the purifying, health-giving sun and air for the sole purpose of preserving the death-dealing carpets and upholstered furniture, which would otherwise become faded.

Dampness enters at night, especially if windows are opened for ventilation, as is the practice in my case both winter and summer, and this is often added to by the use of a washstand in the bedroom. There would not be so much harm in this if the sun were allowed to shine in during the day and the air to circulate.

But although architects and builders make ample provision for sunlight and ventilation, their good intentions are thwarted by overcareful housekeepers, anxious to preserve their carpets, which are the source of so much disease and at the same time add fully a third to their labors. Of the baneful mould stirred up by the sweeper's broom from a carpet, the larger part, probably three-fourths, settles back again or adheres to curtains, walls, pictures and other articles of furniture, to be dislodged by the slightest draught, carried in the air, and inhaled by the occupants of the house.

While writing this I learn from dispatches in the daily papers that acres of carpeting are being laid in the Capitol at Washington, in preparation for the coming session of Congress. After a month of such service as those carpets will see, they will be fit to be taken out and burnt. The same may be said of the carpets used in the other government offices and halls, and even of the carpets of the White House itself.

Constant going and coming, carelessness in the use of spittoons, dropping of fragments of lunch, and the throwing on the floor of anything and everything

17

one wishes to rid himself of—all these practices, indulged in by those who have no personal interest in the protection of the property of the government, help to render the carpet a mass of impurity and an abode of disease.

In the Senate and in the House pages are continually running to and fro, raising clouds of poisonous particles which are none the less deadly for not being seen. If the state of the atmosphere on a busy day in either chamber could be shown by darkening the room and sending a beam of light through it, members would not be surprised at the frequency of headaches and faintness resulting from a session which may last from three or four to twelve, sixteen or twenty hours.

Is it any wonder that so many men ruin their health by devotion to their country's service and their constituents' interests at Washington?

If, at half the expense required for carpets, the Government would construct an open aqueduct, of glass or other material, a foot, at least, in depth and width, extending around the hall under or outside of the gallery, and would keep a constant stream of pure water running during sessions, supplying simply a strip of rubber matting in the aisles to deaden the sound of feet, the air would be more fit to breathe. Plants should also be placed in different parts of the room, and the rubber matting subjected to the daily application of the hose. A change of matting could be provided in order to allow time for cleansing and drying of one set while the other was in use.

And what is true of the chambers of Congress is

18

true of all legislative halls, schools, churches, hospitals, theatres, and also of many offices and workshops, where many people are crowded together, and especially of the composing rooms of newspapers. Floors should be of glass, tiles, marble or other polished stone; if of wood, all joints should be well calked and painted, so that no dust, insects or moisture could find lodgment there.

When this course is pursued as to floors, and all unnecessary drapery and fabrics banished from use, then, and not till then, will the age of man be lengthened, and such diseases as consumption, hay-fever, typhoid fever, and all affections of the respiratory organs will be kept in check.

What clothing we do wear should be, and doubtless will be in the near future (so varied are the resources of this age of invention), of such material as to render impossible the gathering of moisture and consequent mould. Looseness of apparel and sufficient ventilation, so far as is consistent with warmth, will be the important characteristics of our attire.

With our ever-increasing knowledge of the needs of the system in regard to food, our skill in its preparation, our means of caring for the body in sickness, and all the resources of an advancing civilization at our command, there is no reason why our race should not again attain the age reached in Old Testament times.

The power employed to supply electric light and to run elevators and machines in all large buildings in our cities, could be utilized to pump pure water from artesian wells or springs for the supply of such aqueducts as that already described. The purifying power of running water is well known, and by its aid in absorbing and carrying off all atmospheric impurities, the danger of contracting disease by the inhalation of poisonous germs would be greatly diminished.

The condition of our street-cars and railway carriages must be alluded to here. The seats and backs, stuffed as they are, and covered with plush or tapestry, and coming constantly in contact with passengers of every degree of uncleanliness, who may not have changed their clothing for weeks or even months, and who are, perhaps, reeking with unwholesome exhalations from the body—must be filled with impurities.

A car may have been for months at the repair-shops or in the company's sheds, during the dull season, where no sun or fresh air has penetrated. The mouldy condition of its cushions after such a period is something alarming, and when they are subjected to the motion and jolting of travel, and are brushed by the passengers' garments, they send out volumes of disease-promoting particles, to be inhaled by every occupant of the car.

The peculiar discomfort attending a long railway journey, especially in such weather as requires closed windows, is familiar to all, as is the feeling of relief on escaping at last from confinement and breathing a pure atmosphere. Car-seats should be upholstered with some material that is incapable of holding dust or moisture.

Every seamstress, housewife or tailor, who has to do with the making over and alteration of garments that have been worn, knows that the seams and over-

20

laps of seams in those parts that are worn next the person, are clogged with thick layers of mould. The process of ripping, tearing and shaking which the garment undergoes, generally at the hands of several persons at the same time, fills the air with particles of this dead vegetable growth and its spores; these are necessarily inhaled in great quantities by those in the room, and in the talking and laughing that is going on, a portion finds its way through the mouth directly to the throat, and thence to the stomach.

Vigorous exercise in the open air would expel this from the system; but such exercise is rarely or never taken by those who sew for a living, and the poison is suffered to do its deadly work. This is, in my opinion, one of the many causes of typhoid fever.

The case of two young ladies came under my observation not long ago; they made it their business to go about sewing for different families, my own in the number. They were strong, healthy girls, to all appearances, and both came of healthy parents. Yet, within two years of each other, they both sickened and died of typhoid pneumonia, one of them only three months ago. The parents, brothers and sisters of both are still living and apparently in the best of health.

All family clothing to be altered should be taken to pieces in the open air, with the wind blowing from the worker; it should then be carefully brushed on both sides until it is free from all mould and dust, and then hung in the sun and air for four or five days. After that it may be taken into the house and worked upon without fear of ill effects.

21

A very injurious practice, and one that is almost universal, is that of brushing and shaking wearing apparel in the house. In this way the dried particles and spores of fabric-mould and dust and germs from the streets are dislodged and scattered in the air, to be inhaled by the members of the family, or to fall upon carpets and furniture, ready to be set flying again on the next sweeping-day.

All clothing which requires brushing should be taken outside and well away from the house: if no lawn or porch is available it should be taken to the roof, and, while being brushed, should be held in such a way that the wind will carry the dust away from the person brushing it. It should be remembered, moreover, that most garments, such as pants, coats. vests and dresses, need more careful brushing inside than outside; they should be turned and especial attention given to all seams and folds. It is there principally that the mould will be found.

Never sweep or dust a room or brush clothing when a sick person or young child is present, as they are unable to resist the injurious effect of the noxious particles inhaled or to expel them by taking vigorous exercise in the fresh air.

One of the principal causes of malaria and typhoid fever is probably to be found in the construction of suburban, village, and country houses. They are built, in part or wholly, on the top soil, with only a partial excavation or none at all for a cellar. Many so-called first-class cottages are built on stone. brick or wooden underpinning, from twelve to twenty inches high. not a spadeful of earth having been removed from beneath.

Now it is very important that an excavation should be made, from twelve to twenty inches deep, before the foundations of a house are laid, for the reason that the surface soil contains vegetable life that requires the sun and air for its healthy growth. If these are excluded, the growing plants and weeds decay, and all sorts of noxious fungi and seeds of malaria spring into life.

Every summer there is fresh growth and fresh decay. All floors are, as a rule, laid with imperfect joints, and they are very commonly of the cheapest pine boards, which readily absorb moisture and foster the growth of mould. Through the cracks much of the moisture from beneath makes its way into the room, and is both a source of disease in itself and the cause of much mould and mildew in the carpets which it impregnates. In a thickly populated, highly cultivated district, the necessity of excavating before building is so much the greater, because the soil is all the richer in animal and vegetable matter. Not only houses but barns also should have the top soil removed from beneath.

People who go to the country in the summer for their health, often return with the seeds of malaria in their systems, contracted in poorly-built, cellarless houses. With the first cold which they catch, or perhaps at the time of changing their clothes in the fall—when they too often put on underclothing and other garments that have not been properly shaken, brushed and aired—they fall ill and are surprised that they have received so little benefit from their vacation.

The walls, carpets, curtains and stuffed chairs and lounges in a house are generally more or less damp—

23

especially in a room or in a house that has been closed during the summer—when the first fires are started in the fall. Fires are very often built, too, on a cold, rainy day, when doors and windows have to be kept closed.

The evaporation from the drying fabrics, used in the furnishing of nearly every room, can not but be very injurious to health, especially to persons at all inclined to diseases of the throat or lungs. They could more safely be out in a heavy storm, wet to the skin, but inhaling a pure atmosphere, though moist, than in a room where this evaporation of foul and malarial dampness is in progress.

This is the season so trying to all those afflicted with lung or throat diseases—the season, too, that brings with it such epidemics as diphtheria and typhoid fever, to the attacks of which even the strongest and most robust are liable.

It is important to bear in mind that from five to fifteen days are required to dry a house properly in the fall, even with the aid of good fires. If the season is mild, as have been the last two and the present up to the time of writing (Dec. 15), fires are either kept very low or not built at all, excepting in one or two rooms most in use. This is particularly true of the lower and middle class families, and in such households two, three or even four months elapse before the houses are really dry and habitable. Meanwhile the occupants breathe an atmosphere unwholesome in the extreme, and countless cases of typhoid fever, diphtheria or other complaints of the throat or lungs, call for the doctor's assistance.

The unusually wet summer which this vicinity

24

experienced the present year, combined with the mild fall and tardiness of people in building fires, explains the prevalence of the above-named epidemics. Rooms in hotels are, above all others, imperfectly cared for. Mould and moisture are allowed to remain year in and year out, in many cases, carpets are not removed for long periods, and while unoccupied, the rooms are kept closed and dark. Sick patients have small chance of recovery if confined in such rooms as these, even with the best of medical attendance.

Rapid as is our advance in the sciences and arts, we are still far behind many of the nations of antiquity in the matter of wholesome dwellings. Excavations in Italy, Greece, Asia Minor and elsewhere show that even the most ordinary houses had floors of marble or of some other stone, and their construction throughout was commonly of stone, brick or other masonry. In the main room or hall, or sometimes in an inner court, was its fountain of running water. But in this age when we can put steam into harness and make it do our work, we suffer for the want of fresh air and fresh water, and live in decaying wooden houses with carpeted floors.

The tolerance of an impure or insufficient water-supply is inexcusable, when, with the improved appliances for doing it, artesian wells can be so easily sunk or water obtained, by pumping, if necessary, from some distant spring. With wells sunk to such a depth as to render impossible contamination from surface drainage, we should be unaffected by any such accident as that which recently left the city of Brooklyn unprovided with water.

25

HAY-FEVER.

For a long time physicians have been convinced that hay-fever is caused by taking into the nostrils and air-passages certain forms of vegetable matter, either whole or in a minute state of subdivision. The disease makes its appearance in late summer and early autumn, and is very commonly ascribed to the inhalation of the pollen of the rag-weed—an explanation, however, which but inadequately accounts for the wide prevalence of the distemper. It is caused by taking into the nostrils bits of mould in its living state, or its spores, in the manner already described. Inflammation of the mucous membrane, or what is known as a cold in the head, may have been caused beforehand by the autumn dampness of the house, and the moisture of the nasal passages offers a suitable lodgment for the particles of mould or the spores which have ripened in the fall, as all plants and weeds ripen at that time.

Just as a cutting from a willow or a geranium will take root and grow on being stuck into suitable soil, so the tiny fibres of mould retain their life when they are transplanted to the mucus of the nasal passages. The spores, too, which are inhaled, germinate and grow. But in the same way that a plant, on being transplanted to a flower-pot containing only a shallow layer of soil, will wither and die as soon as it has extracted the nutriment and moisture from this meagre supply of earth, so the growth of mould in the nostrils, as it does not take root in the animal tissue itself, ceases as soon as the supply of moisture fails, either from natural causes or through artificial means. And as any other dust or foreign matter is readily expelled from the nose, so the dead and dry mould is driven out and the hay-fever is gone. Its termination may be caused by a season of dry weather, by the removal of the patient to a dry climate, or by his sitting over a warm fire in a dry room. The one thing needful is to allay the inflammation of the mucous membrane and restore it to its healthy condition.

The mould growing on carpets and other fabrics is the prime cause of hay-fever, and the peculiar irritation and tickling sensation felt in the nasal passages is due to the passing of air over the strands or filaments of growing mould, which are thus swayed back and forth, while some are detached and driven further in, causing the sufferer a fit of sneezing. As a speck of dust in the eye irritates and inflames that sensitive organ, and as a hair in the throat causes great annoyance, so the smallest bit of mould-fibre in the nose tends to irritate and inflame the sensitive membrane with which it comes in contact. This very irritation provokes the flow of mucus, or running at the nose, which is essential to the life and growth of the mould; so that the distemper provides for its own continuance and baffles all attempts at a cure, unless treated with the full understanding of the prime importance of a dry, pure atmosphere.

Probably no victim of hay-fever ever had his case brought into such prominence as the late Henry Ward

27

Beecher. Others may have suffered as much or more from the complaint, but their sufferings were known only to their friends and their physicians. The whole country has heard of Henry Ward Beecher's annual attacks of hay-fever, so that any allusion to his experience will form the best possible illustration and proof of the foregoing.

As is well known, Mr. Beecher was in the habit of spending the summer on his farm or in travel, while his Brooklyn residence was either closed entirely or left in charge of servants. In late summer or early fall, before settling down for the winter, he would come to the city occasionally to conduct religious services at his church. Word was probably sent at these times by mail to the chambermaid to have his room in order on a certain date. Now, it was this opening and sweeping of a long unused room that gave him his annual hay-fever. The danger encountered in entering and sleeping in such a room will be apparent to the reader from what has gone before. Dampness, drying fabrics, dislodged and floating mould particles, torn up by the roots like weeds pulled up in the garden and still alive—all the conditions necessary for hay-fever were present.

It is probable, too, that the occupant of this imperfectly prepared chamber was often obliged to sleep between sheets that had not been thoroughly dried, and to use towels in the morning that had lain in the closet all summer, full of living mould and its spores. The unwholesome dampness of everything in the house must have been all the greater from its nearness to the water. The preliminary cold in the head was easily caught, and then followed hay-fever from the inhalation of floating particles and spores of mould.

If, instead of going to his own house, Mr. Beecher went to that of a friend, he was probably given the guest-chamber—more properly the death-chamber, damp and mouldy as this apartment always is from infrequent use—and here the same causes were at work as at his own home. His sedentary habits, be it further observed, rendered him more liable to the attacks of the malady than one engaged in active work in the open air, and, as a natural consequence of his mode of life, he could only recover his customary health after his house had been thoroughly dried and aired in the manner explained above.

What was true of Henry Ward Beecher is true of every victim of hay-fever. The same causes produce the same results, wherever and whenever they are present. The reason why some people have the disease, and others who are exposed to the same danger escape, is found in the stronger constitutions or habit of regular exercise of the latter class. There is little danger of incurring the epidemic in rooms that are occupied throughout the year, and women, whose duties or habits keep them in well-warmed rooms, are not in much danger.

DIPHTHERIA.

This is also a disease which can be traced entirely to fabric mould and the poisonous evaporation from drying carpets, drapery and other fabrics, principally in the fall and winter seasons. It broke out in my own city, Beverly, about the first of last September, and spread with such rapidity that the schools were closed for a short time. A few cases still linger as I write this—more than three months after its outbreak. During July and August of this year there were thirty days of rainy weather—that is, thirty days on which some rain fell—in Beverly and about Philadelphia. So much damp weather necessarily had an effect on dwellings, and the more so that it came at a season when no fires were kept and nothing prevented the free entrance of the moisture into the house. As a consequence, all carpets and stored fabrics were more heavily laden than usual with deadly mould and damp when fires were started in the fall.

The chief cause of diphtheritic outbreaks is found in the annual or semi-annual house-cleaning, conscientiously performed by all good housekeepers. No one inclined to lung or throat diseases, and no young children, however strong and healthy, should remain in one of our carpeted dwellings during this operation, or return to it until fifteen or twenty days after its completion. I believe that house-cleaning unintelligently performed, as is commonly the case, is one of the chief

30

causes of many diseases. Typhoid and scarlet fever, diphtheria, and all affections of the respiratory organs, may often be found to have sprung, directly or indirectly, from this cause. A brief consideration of the ordinary methods used in cleaning a house will make this plain.

When carpets are taken up, clouds of poisonous dust, of the nature already described, fill the rooms, and the particles lodge in every crack and on all pieces of furniture, curtains, pictures, and other objects. No duster can find all this poisonous powdery organic matter, and even if it did, the particles would only be driven into the air and left to settle back again in a short time. For days afterward this dust is likely to be stirred up and sent flying with the casual brushing of a garment or a chance draught of air. The carpets are themselves taken out and beaten in what seems a very thorough manner, though it is hardly probable that more than a quarter of the adherent mould, organic matter, and dust is ever driven out. The moisture that has settled in every thread of the carpet is dried little, if any, by this beating and shaking. Meanwhile floors are scrubbed, and when the surface of the boards appears dry, the carpets are replaced— often on the same day as their removal.

From five to ten days are, in reality, necessary for the thorough evaporation of the water that has been sucked up by every crack and loose joint in the long-seasoned boards. Spaces of an eighth to a half of an inch are very common between the boards of an ordinary floor, and these are filled with dust and mould and other impurities which absorb water readily and

31

give it up slowly. The injurious effect of shutting in all this moisture by the laying of carpets may be imagined. One can scarcely conceive anything more nauseating and unwholesome for a person whose respiratory organs are at all affected than the fumes from a soft-wood floor just scrubbed with soap and water. It means a rich harvest of unwholesome organic matter before the next cleaning day comes— in six months or six years, according as the mistress of the house is a careful or a negligent housekeeper.

The injurious evaporation from damp carpets and from damp floors through the carpets, begins in earnest when fires are made, and the winter is nearly over before it stops, even in the best heated houses. October, November and December are the months most fraught with danger in our latitude, the house from top to bottom being filled with poisonous exhalations and infected with disease-germs. Is it any wonder that this is the season of typhoid pneumonia, scarlet fever, diphtheria and kindred diseases!

Statistics show that diphtheria is mostly confined to children from five to fifteen years of age. This is just the age when they are most in the habit of playing, running and rolling on the floor, especially when it is unpleasant outside, laughing, crying or talking the while, and breathing more often through the mouth than through the nose. It is unavoidable that they should take in with every breath quantities of poisonous organic particles which, in their romping and games, they have stirred up from the carpet. These injurious particles settle in the throat, irri-

32

tating the sensitive mucous membrane with which it as well as the nostrils, is lined, until inflammation and soreness ensue, offering just the kind of soil needed by the inhaled mould and spores for their most rapid and abundant growth. The result is unhappily too well known to require description.

The membrane of a child's throat, as, in fact, its skin in general, is more tender and sensitive than an adult's; and as children spend much of the time playing on the floor, and are always, even when erect, obliged to breathe the lower strata of air, they necessarily inhale, and are affected by, the dust of the carpet. Grown people breathe less frequently through the mouth, and their respiration is in a stratum farther removed from the floor.

When a case of diphtheria becomes malignant, the patient is a source of contagion, as every breath or cough sends out from the diseased throat particles of the deadly fungous growth that is eating away the very life of the child. Other children, whose throats are nearly always ripe for the infection, on inhaling these particles, become themselves liable to attacks of the disease. Girls are more subject to diphtheria than boys, for the reason that they spend more of their time playing in the house.

A lucifer match may start a fire that eventually destroys a hundred houses in a thickly-populated district. The original cause of the conflagration was the one small match and the one building set fire by it; the remaining ninety-nine caught the fire from the first house. So with diphtheria and other contagious diseases, one careless and untidy housekeeper, or

33

the want of sanitary laws, may, through negligence or ignorance, provoke a contagious disease that will spread far and wide, killing hundreds in its course.

A word must be said here about the importance of properly salting a child's food. Cooks often send the dishes to the table insufficiently seasoned, leaving that to be done according to each one's taste. Grown people easily supply what is lacking, but very young children are unable to attend to their own wants and are often neglected by their mothers or nurses. In consequence, they eat much of their food unsalted or salted insufficiently, and their systems are, as a result, rendered more open to the attacks of many forms of disease, either through contagion or otherwise.

Neither insects, mould or mildew will grow on anything that contains an appreciable amount of salt, and if the human system is well supplied in this respect, it has a safeguard against many diseases. Salt should, however, be taken in moderation, as too much is decidedly injurious, though in a manner quite different from too little.

If carpets were discarded, house-cleaning would lose its terrors, and with proper care in the brushing, drying and airing of all fabrics used in the house or on the person, diphtheria would be unknown, as well as scarlet and typhoid fevers, which are probably caused by the same poisonous organic matter that gives rise to diphtheria.

Adults, more than children, are subject to typhoid fever, for in their case the lining of the throat is less tender and less easily irritated by the inhaled mould particles. These pass down to the stomach, either directly or in eating and drinking, and the poison enters the system through the digestive organs.

In many farm-houses and village cottages, the best room, or parlor, is only opened on the occasion of a party, or in honor of some especially favored guests. Accordingly, the room is rarely heated, and when a fire is made, the state of the atmosphere, from the drying of accumulated moisture and mould, beggars description. Whole days of continued fires and ample ventilation would be required to render the room habitable. The guests, then, take their lives in their hands when they enter such a parlor, and their entertainers, all unconscious, share the peril with them.

CONSUMPTION.

Consumption also is to be ascribed to the mode of life prevailing among civilized nations. No physicians have ever yet been able to cope with the disease, for the reason that none have perceived the all-important influence exerted upon our respiratory system by the nature of the atmosphere which we breathe. They have disregarded the wonderful vitality and insidious action of mould, mildew and other forms of microscopic fungi, which always flourish in houses furnished with carpets and upholstery. The cause of consumption does not differ in kind from that already given for hay-fever and diphtheria. The same agents operate in destroying the lung tissues as those which irritate and clog the nasal passages and bronchial tubes.

I do not believe that consumption, or any other disease of the respiratory organs, is transmitted by inheritance. The lungs are as perfect in their structure at birth as the hands, feet, or any other part of the body. To be sure, the lungs of a child born of consumptive parents may be less robust than those of the offspring of a perfectly healthy man and woman. This weakness could, however, be outgrown under the proper conditions. The disease itself invades the lungs only during life, and, of course, the stronger the constitution is originally, the more readily will it be repelled.

Young and growing children are rarely attacked

by consumption, because they keep their lungs sound by abundant exercise, often in the open air, running, jumping, shouting and laughing, and keeping the sponge-like pores of their lungs clear by vigorous respiration. Indolence and inactivity are largely to blame for the disease.

Indians, when brought into a state of civilization and led to adopt the domestic customs of white men, living in carpeted houses instead of roaming the woods and prairies, become liable to its attacks before one generation has passed.

A laborer in the fields, on catching cold, suffers some little annoyance from the flow of thin, watery mucus from the nose; but as he keeps on working and taking in copious draughts of fresh air, this emission soon ceases and the nasal passages are restored to their normal condition. But let the same man be confined in a carpeted room, on catching such a cold, and he will not get rid of it so easily. The mucus, which had before run as water, will begin to thicken from the admixture of mould-fragments, and spores, lint and dust; the air passages, from the nose down to the throat, and perhaps to the lungs themselves, will become clogged, and he may find that his cold has developed into catarrh, bronchitis, asthma, or even consumption, so rapid is the advance made by the parasitic fungous growth which has spread from the carpet to his respiratory system. The mould has obeyed the law of its nature, and gained a foothold wherever the requisite conditions were offered.

According to the part which it attacks, the victim has bronchitis, catarrh, hay-fever, deafness, asthma or consumption, and if no vigorous opposition is made to its ravages, on the part of the patient, death ensues. The wheezing sound peculiar to the breathing of asthmatic sufferers is due to the passage of air through air-tubes clogged with the deadly fungous growth.

However many opposing theories may be maintained in explanation of consumption and diseases of the throat, the facts of the case remain unaltered. At death the air-tubes collapse, as does a balloon on the escape of the inflating gas, and no *post-mortem* examination of throat and lungs can be satisfactory, as throwing light on the exact nature of their maladies. But if careful study could be made of the respiratory system of a living consumptive or asthmatic patient, the truth of the above statements would be confirmed.

Those who would be sound in wind as well as limb must not shrink from hard work or exercise in the open air. No system of parlor-gymnastics will suffice. The very exercise which some people take so conscientiously in their chambers only adds to their throat or lung troubles, through the stirring up of particles of mould from the carpet or furniture, or even from the clothes of the one exercising, causing the speedy vitiation of the confined atmosphere of the room.

Gladstone would never have attained to such a hale and hearty old age, if he had brought his logs into the parlor to chop. No exercise can be more wholesome than the vigorous handling of an axe. With every stroke the air is driven through the lungs in a miniature cyclone, clearing away obstructions and imparting vigor to the tissues.

38

Last year at this time there was a great flocking of consumptives to Berlin, on the announcement of Professor Koch's consumption-cure. The newly discoverd lymph was made to fill columns in every newspaper, and large head-lines called the reader's attention to the wonderful discovery. Jenner and his vaccine were for the time in danger of being eclipsed by the more modern discoverer and his lymph. Side by side with the announcement of this medical discovery appeared, with equally large head-lines, dispatches from South Dakota, describing the uprising of the Indians and their ghost-dances.

I often thought, while these two topics were furnishing so much copy for the press, that if sufferers from consumption would turn their faces toward the West, instead of toward Berlin, and would betake themselves to the scenes of these hair-raising pranks of the Indians, they would stand a better chance of being benefited. One raid of the red-skins and a life-and-death chase of four or five miles would have done more to clear and invigorate their lungs—provided the disease had not been too long neglected—than all the lymph or other medicines ever made.

Some will say, in opposition to the preceding pages, that consumption and other diseases of the respiratory system are common among many races that never saw a carpet or any upholstered furniture, and have never even had a dwelling worthy of the name. This is very true; but some clothing is worn by nearly every race on the globe, and mould and other varieties of fungus are found everywhere. The inhabitants of the torrid zone are, it must be noted, generally indolent, and often live simply on what nature offers them, without any effort on their part.

When I was in the army I slept on a rude bed made of hay confined by a rough frame of boards. During damp weather the hay was always mouldy and needed to be changed often on that account. All clothing and blankets were necessarily very damp and mouldy; but from this circumstance I suffered no injury, owing to the active life imposed upon me. Brisk exercise and the vigorous respiration accompanying such exercise kept my lungs in a clear and healthy condition. If I had led an indolent life at this time, such as is common among many savage races. I should have been soon affected.

Sailors scarcely ever take a cold while at sea, and if they do, it is soon cured, the thin, watery discharges drying up of their own accord. But if a sailor takes cold when on shore. and lodging in an ordinary carpeted dwelling, the case is different: mould, lint and dust are present to change the watery mucus into a thick clogging mass which seriously retards the restoration of the membrane to its healthy condition. The sailor never has such an experience at sea—a fact which proves conclusively the existence of certain dangerous, unwholesome conditions in our dwellings.

Air, water and salt play each an important part in what may be called the earth's circulatory system. Each is necessary for the purification of the other two, and all are needed for the sound and wholesome condition of the earth. A shower purifies the atmosphere: free circulation of air helps to keep water from stagnating; salt renders wholesome the vast oceans of water.

into which the rivers are constantly emptying their floods for purification; and the winds that come to us from over salt water bring health and strength with them. The ebb and flow of the tides is a kind of vast respiration, carrying off impurities and bringing back renewed health and vigor. The important office which the unceasing activity of air and water and the purifying action of salt are intended to fill in the human system, is thus indicated to us in the operations of nature.

INFLUENZA.

The prevalence throughout the Middle and New England States, at the present writing, of a distemper, almost epidemic in its nature, and called by many the *grippe*, but in reality a kind of influenza or catarrhal fever, makes it especially appropriate to say a few words concerning the latter disease.

The causes which led to it are of the same nature as those already discussed. The unusual number of moist and rainy days during the last summer left our dwellings in a very damp and unwholesome condition when the time came for starting fires in the fall—or, rather, the time when we ordinarily start them. Consequently the accumulated dampness of the past summer has not been dried, and people have continued to live and breathe and sleep in rooms which, in ordinary years, are in a tolerably suitable condition for habitation by this time. They do not realize the damp, mould-infected condition of their dwellings, or if they do, are not alive to the danger which they incur in living in them.

The present distemper is more prevalent and more fatal among the poorer classes than among the wealthy and well-to-do professional and business classes, differing in this respect from the *grippe*, which attacked the higher orders of society with perhaps even greater violence than the lower.

The explanation of this is simple: the poorer families have profited by the warm weather to save fuel,

42

and have kept up no regular fires, as they are ordinarily compelled to do from the coldness of the weather. In the houses of the wealthy, on the other hand, furnaces or steam boilers have been going much as usual, being, however, kept down to the lowest point.

The condition of the imperfectly dried or wholly undried dwellings has been going from bad to worse, as the dampness clinging to carpets, bedding, garments and all fabrics and absorbent material has generated poison, and caused the growth of mould to an ever increasing extent.

A lawn is hardly ever so parched and dry that a shower will not start the green grass almost immediately. So the parasitic fungous growth that lurks in some corner of every one's respiratory system may be kept under control year after year by sufficient exercise and the avoidance of a malarial atmosphere. But when once the proper conditions of damp, foul air and floating impurities are present, the reinvigoration and spread of this unwholesome growth is rapid, especially in the case of the aged and the indolent, until the very bottom of the lungs is reached, if the danger is not vigorously combated, and the patient is choked and strangled to death.

If a swamp or stagnant pool, although open to sun and air and every breeze that blows, is the cause of so much malaria, and is so fraught with danger to human life and health, how much more poisonous must be the evaporation from the stagnant moisture in our soiled carpets and other textile fabrics, in close rooms which may seldom or never receive a thorough airing?

The action of moisture on dyed stuffs ought to re-

43

ceive some attention. Many poisonous aniline dyes are used in carpets, upholstery, wearing apparel and fabrics generally. It is even impossible to tell how deadly poisonous some dyes may be. Their preparation and use are, in many cases, a secret known only to the manufacturer, and it may well be that many a carpet or piece of wearing apparel gives off, when damp, unwholesome or deadly fumes. We at least know that the smell of wet clothing is often sickening in the extreme.

All that is necessary in order to drive away the influenza is a hot fire in furnace or stoves, kept up for ten or even twenty days. Let each room in turn be converted into an oven, or something resembling one, while the members of the family seek another part of the house. By this process the house will be speedily and surely rid of all undue moisture.

The change to colder weather, which has set in as this goes to press, will doubtless cause people to follow, in a measure, the above directions. They will keep up good fires and thus drive out the cause of the recent illness.

To sum up the matter briefly, influenza is a complaint similar to hay-fever and is to be ascribed to the same cause. But the fungous growth which operates in both instances is more irritating and troublesome in the former, because of the lateness of the season: no Indian Summer days intervene to dry up the flow of mucus and relieve the patient. In the case of *grippe*, however, an exactly contrary condition of atmosphere and temperature is to be noted.

On that Christmas day of 1889, which just pre-

ceded the outbreak of the epidemic that was to visit
all sections of our country like a plague, the air was
as dry and balmy as on a beautiful May day. On that
and a number of preceding days the mercury went up
to 75° or more, even in the shade, and what wind there
was came gently from the south, with no rain or
dampness whatever.

People congratulated one another on having such
fine weather for the holidays, little thinking that just
this fine weather was bringing to life countless millions
of microscopic insect germs, prematurely hatched,
which were being inhaled with every breath, and
that before three days should pass a large proportion of the inhabitants of the country, north, east
and west, would be stricken with an epidemic,
the like of which had never before been known
among us, carrying death in its course and the
undermining of many a strong constitution. All
agreed that such a balmy Christmas day had never
been seen before, and the oldest inhabitant could not
recall another instance of such a continued high temperature at that time of year.

If on some Fourth of July, when the crops are
well started in the fields, the mercury should suddenly
drop to 30°, people might congratulate themselves on
their welcome escape from the oppressive heat usually
characteristic of the day. But on learning, immediately afterward, of the total ruin of crops wherever
this temperature had prevailed, they would cease to
look upon such unseasonable weather as a blessing.
Yet it would be no more unseasonable than the
weather which we experienced on Christmas, 1889.

While a sudden change to cold, dry weather will necessitate the building of fires, and thus hasten the drying of moisture and the consequent cure of influenza, it is otherwise with the *grippe.* From five to fifteen days, and in some cases weeks, are required to expel the disease from the system, even with the mercury below zero. The germs in the open air are quickly killed, but those in the system and in warm dwellings more slowly. A very heavy death-rate from *grippe* was noted on some of the coldest days.

It should be borne in mind that this disease owes its origin to insect germs which no process of drying can kill, while the influenza is to be ascribed to a vegetable or fungous growth which can be subdued by drying.

PULMONARY DISEASES IN CATTLE.

These are caused by mouldy hay, grain or other fodder. In my boyhood it was one of my tasks, during the harvesting of the hay, to follow the mowers and stir and spread the freshly-cut grass for drying. This work was always begun early in the morning in order that the hay might be dry enough to stack or put in the barn before night. Not infrequently a sudden shower defeated our plans, and the hay had to go through the same drying process the next morning. On these occasions I always found it mildewed underneath, when it was first shaken out, and if showers came several days in succession, the same thick mould was always found on the following morning on the under side of the hay. Frequent stirring and exposure to the sun and air would dry it, but the mould, too, was dried, and adhered to the hay. Consequently, when the hay was finally put away in the mow, it was often mixed with large quantities of dry mould. Hay is never entirely free from mould, and I have often been nearly suffocated by it when "mowing away"—to use a farmer's expression—so that, though strong and robust, I was frequently compelled to seek the nearest opening for fresh air.

47

Stalls for horses and cattle, especially in country barns, are commonly so built that the animals stand facing the barn floor. When hay is thrown down from the mow, preparatory to feeding, suffocating clouds of powdery dust, composed of pulverized mould and its spores, rise into the air. The men can relieve themselves by going to the door for a breath of fresh air, but the animals must stand and endure it.

Very often the threshing and winnowing of grain is done on the barn floor, with horses, and perhaps cows also, standing in their stalls, with their noses in the thickest of the chaff and dust which fill the building in such volumes as to render it often impossible to see from one side to the other. I was, of course, too young to understand the injury our horses and cattle must suffer in breathing such an atmosphere, but it is clear to me now why pleuro-pneumonia is so common among cattle on a farm. It was our custom, moreover, to feed to the cattle corn-stalks which were kept piled in shocks near the barn, exposed to the weather. Toward the middle of a shock the stalks were found more and more thickly covered with blue mould and were fed to the cattle in that condition; this could not be otherwise than harmful to the cattle.

Mould grows everywhere under proper conditions, even in the open fields and in woods. Out on the western prairies, where so many cattle die in severe winters, they are forced to subsist almost entirely, in the winter season, on mildewed tufts of grass, which reveal on the lower side a thick coating of blue mould. The same condition is observed under the bunches of forest leaves in winter, when the snow is scraped off and the leaves turned up to the light. On this kind of provender cattle on the plains have to depend during much of the winter, while cows and horses that are sheltered in barns are often fed on mouldy hay or grain. The consequent prevalence of pleuro-pneumonia among cattle is not to be wondered at.

It was formerly the custom and may be still, in shipping eggs from the West to New York or other eastern cities, to pack them in oats, filling barrels in this way. I remember a grocer of my acquaintance who had such a large retail business that he kept five horses for running his delivery wagons. On one occasion his hostler fed these horses with the mouldy oats which had come from the West as packing. The result was the death of four of the animals within a week, while the fifth was made so sick that he had to be sold as useless.

Another reason why pleuro-pneumonia, glanders and other diseases are so common among cattle is that their systems have not sufficient salt to counteract the poisons which they take in through stomach or lungs. When fed on dry winter fodder, cattle manifest an eager craving for salt, and on being shown a handful, will make the most violent efforts to get at it. It may be given them mixed with their hay or grain, or thrown to them in the yard. When it is fed to them out of doors, the dull, stupid looking cows and oxen will display all the life and alertness of a mettlesome horse, so great is their desire for salt, and if the one feeding them would escape being trampled upon, he must throw out a handful and run, and then another, repeating the process until the animals have lost the edge of their appetite.

Sheep show an equal fondness for salt, and it should be given to them, as well as to the cattle and horses, as often as once a week. On our farm it was the custom to do this every Sunday morning. The action of salt on all vegetation, and hence on such fungi as mould and mildew, is to kill it. Its presence always checks the growth of unwholesome vegetable and animal parasites.

THE MODEL DWELLING-HOUSE AND ITS FURNITURE.

Whether built of stone, brick or wood, dwelling-houses should be of such perfect construction that no external moisture or air could enter, except by the will of the occupants, through doors, windows, or other means of ventilation expressly provided. Inside walls should be finished in hard wood or plaster and painted, while, from time to time, a coating of lacquer should be applied, to preserve the perfect smoothness of the surface. No paper should ever be allowed on the walls.

It is probable that the time will come when some kind of opaque glass, porcelain, or other glazed earthenware will be generally used in interior house construction, both for walls and floors. Even now we sometimes see a room whose walls are of glazed tiles to the height of three or four feet, the remainder being of hard-finished plaster. Tiles for floors are constantly coming more into favor. The material used should be in all cases of such a nature that no moisture, dust or vermin could penetrate or find lodgment there.

If, from considerations of expense or otherwise, wood is used in the construction of floors or walls, every joint should be perfect, if possible, and the imperfect joints calked as carefully as a ship's seams. Then, with the application of paint and a frequent coating of the hardest kind of varnish, the surface can be kept sufficiently smooth.

To those who may object that the style of house building described above is too expensive, it may be stated that a coat of whitewash, frequently applied, protects the walls from harmful germs and mould. No fine, hard-finished plaster can be more wholesome than a fresh coat of whitewash.

Rugs or matting of some kind could be laid here and there, if desired, and these should be taken out daily, and, by beating, shaking and brushing, thoroughly cleansed of all dust. Every article of fabric not in daily use should be kept until needed in a room especially designed for such storage, where the entrance of mould, moisture, dust, moths or other insects would be reduced.

The seats and backs of chairs, lounges and sofas should be of wicker-work, cane, prepared leather, or other material which is impervious to disease-germs, moisture or the finest particles of dust. Nickel-plated wire could well be employed in making seats of all kinds, supplemented, if necessary for comfort, by nickel-plated springs. No moisture or dust could settle in the material of chairs and sofas thus constructed, the nickel would not rust or tarnish, and it could be cleaned, when necessary, as easily as a piece of glass or earthenware.

Window shades should be made of some kind of fancy wire gauze, or other material equally incapable of harboring dust, mildew or moisture.

Growing plants should find a place in the living-rooms of every house, especially during the seasons when the weather is not sufficiently warm and dry to allow the free admission of outside air. At the same time some provision should be made for cleansing the air by means of pure water. Large sponges saturated with water might be hung from the ceiling, enclosed in nickel-plated wire cages with a water receptacle attached, to catch any drippings from the sponges. These sponges should be frequently cleansed in boiling water, and then hung up again, filled with a fresh supply of clear, pure water. The method of suspension could be that employed in hanging lamps, and the sponges could be raised and lowered in the same way.

In bed-rooms and other apartments not occupied during the day, the windows should be opened in clear, bright weather, shades should be rolled up, and air and sun allowed free entrance.

When the above rules are observed in domestic life, our dwellings will be free from disease-germs, but not before. There will be no mould-growing carpets or useless draperies to bring on periodical illness and increase, by fully one-third, the cares and duties of the housekeeper. Every breath drawn will be fresh and pure, headaches and other maladies caused by a vitiated atmosphere will be unknown, and cases of hay-fever, typhoid fever, diphtheria, catarrh, and other similar complaints will be reduced to a minimum.

The objection may be made that without carpets on the floors our dwellings would in winter be much like ice-houses. But it is a well-known fact that marble, stone, glass and tiles retain the heat, after being once well warmed, much longer than even the thickest

53

woolen carpet or any other material which is of a porous nature. The floors of hotel corridors and other large buildings illustrate this.

A marble or stone floor carries a reserve supply of heat and helps to keep the temperature uniform. In the summer, too, such a floor is preferable to any carpet, being much cooler. But its great point of superiority is, of course, its freedom from injurious germs, whether animal or vegetable. The danger of such germs in the clothing which we wear remains to be guarded against; but after we once well understand the nature of this danger, we can reduce it to a minimum by exercising reasonable care.

THE MODEL SICK-ROOM.

It should contain the least possible amount of fabric of any kind. Sheets and pillow-cases should be of the finest linen, or, if possible, of some non-absorbent material. No fabric having much lint or long fibres should be tolerated in either quilts or blankets. Over and around the bed, on a frame designed for the purpose, should be spread a fine silk gauze or veil, to keep out, as far as possible, any floating dust, mould or bits of lint from clothing. The air would thus be filtered before reaching the patient.

Invalids need the very purest air, having no strength to spend in resisting the injurious action of atmospheric impurities, while, as a matter of fact, the air of a sick room is generally impure almost to suffocation. Sun and fresh air are excluded, the room is more than usually well provided with every kind of furnishing that is soft to the touch, but the lurking-place of disease and death, and the going and coming of nurses, doctors and friends keep the atmosphere thick with all kinds of noxious germs and dust. It is a wonder that invalids ever recover with so much to contend against.

Not only with the sick, but also with very young children, from the time of birth till they are able to run and exercise, great care should be taken to provide pure air. In our ordinary carpeted houses it would be well to place a gauze canopy, like the one described above, over the baby's crib, when he is in it, and, if practicable, a similar air-filter should be over his face at other times.

Very young babies are too weak to resist the harmful influence of atmospheric impurities and it is no cause for surprise that half of them die under the age of five, from croup and stomach troubles, largely caused by the inhalation or swallowing of noxious germs floating in the air.

EXPERIMENTS AND TESTS.

Simple experiments, in proof of much that has been written in the foregoing pages, can easily be made by any one. To render visible the microscopic particles and spores of mould present in any carpeted room, close the shutters on a bright day, with the exception of a narrow crack through which the sun may shine. Then if the narrow sunbeam thus admitted is not already seen to be full of floating particles, it may be rendered so by tapping the carpet with the foot or shaking any garment that may be at hand. Dense clouds of what is commonly called simply dust will appear to add substance to the beam of light, and at the same time minute insects or animalcules may be seen to dart and flash in the light, as a fly darts this way and that in the air. The air in our houses, if kept within certain limits of temperature, is always thickly populated with this microscopic animal life, as is also the air outside at certain seasons.

The animalcules are taken into the system in great numbers with every breath. When decomposing animal or vegetable matter is exposed to the air, they settle upon it and there grow and multiply until, in some cases, their presence becomes manifest to the naked eye.

These animal parasites also prey on living matter, whether in animals or in human beings, especially if there is anything like gangrene or decay to attract them. But if the body is entirely free from everything in the

57

nature of decay, they are harmless, and, indeed, many varieties of this microscopic population of the atmosphere may be positively beneficial, acting as scavengers or serving some other good end in our Creator's design, which is not yet understood by us.

Some may assert that their dwellings are free from the kind of matter described above, and that the test mentioned gives no indication of its presence. The fault, then, lies in the manner of conducting the experiment; perhaps the room has not been sufficiently darkened, or the ray of light has been too feeble. Instead of a sunbeam, a strong ray of light from a dark lantern will serve the purpose nearly as well.

Every form of animal life reproduces its kind by means of an egg, and every vegetable or plant by means of a seed. The vegetable may also, in many cases, send out what may be called an extension of itself, either by its root, as in the potato, or by a cutting, as in the geranium. So all varieties of fungi, whether large, as mushrooms, or small, as mould and mildew, are really plants; and like plants they produce seeds which drop on the ground, or on the floor, or are taken up into the air and carried hither and thither, before finding a resting place suitable for germination.

An atmosphere that is thoroughly dry and wholesome for human beings is fatal to the life and growth of fungi; the mould dies as the grass would die on a parched hill-side, and the spores and dead fragments are then caught up into the air, on being dislodged, and are scattered through the room or throughout the house, ready to spring to life again on meeting with favorable conditions. Unwholesome moisture, dark-

58

ness and confinement, together with sufficient warmth, are the prime requisites for germination and growth.

After wiping the face with a towel and hanging up the latter for the night, we find it, in the morning, sustaining a growth of mould, as may be seen by gently shaking it in the path of a ray of sunlight. The finest handkerchiefs and other linen, after being packed away a day or a week, are found, on careful examination, in the same mouldy condition.

On frosty days the breath is plainly visible for a distance of a foot or more from the face. It is drawn in from the same distance, and consequently we must inhale many of the floating particles that come within this limit. The nasal passages form a kind of filter for the air, it is true, but not all foreign particles are strained out before reaching the throat and lungs. Many people, too, and children, as a rule, breathe often through the mouth, while the habit of breathing through the mouth in sleep is very common. Every movement in bed shakes a portion of the mould-particles from the bed-clothes into the air. My room is often darkened in the morning to keep out the flies, so that, when a beam of light happens to strike the bed, I have noticed how slight a motion on my part would send clouds of particles into the air. A thump on the sheet produces a very marked effect.

Another experiment may be performed by causing some one to disrobe in a box-like compartment, about four feet square and six feet high. If the room is darkened during this operation, and a ray of light directed over the top of the box, clouds of dust, lint, broken fragments of mould and its spores will be seen to issue from it and stream upward, like smoke from a chimney.

An apple, pear, nut or other fruit or vegetable that has a worm-hole or crack of any kind in its substance, soon becomes mouldy at the core; confined, moist air and darkness are all that are needed to produce the fungous growth, whether in the human system or elsewhere.

When a garment or other piece of fabric, or a carpet has once harbored any mould, it can not be entirely freed from it any more easily than a garden overrun with weeds can be entirely rid of them and their seeds. A little carelessness or neglect will easily prove this.

A black broadcloth suit or a velvet cloak that has been packed away during the summer, if shaken in the path of a sunbeam in a darkened room, will give out what appears to be white dust, but is, in reality, fragments and seeds of mould, while mixed with it may be found some particles of lint from the garment itself. That the dust is not all lint, is shown by the color.

Women, while performing such housework as sweeping and dusting, should wear fine gauze veils, as they often do in the street; these should also be worn on railway journeys, and it would be well if men could wear them also.

The constant presence of moisture in our clothing, even when it does not give the slightest indication of being damp, may be proved by standing near a hot fire, when the dampness may be felt coming out and seen in the form of steam.

It is the custom in some families to have the carpets taken up, beaten and stored in the spring, and in the fall they are considered all ready for replacing on the floors. They are, however, much less fit for use than if they had not been taken up at all. Darkness and confinement, with the inevitable dampness present in the carpets at all times, have caused a rich growth of mould, necessitating a thorough re-shaking and exposure to the sun and air.

The presence of moisture in a room and its furnishings at all times can be proved by heating the room to a temperature higher than that outside. Immediately the window panes act as condensers, and are soon covered on the inside with vapor, as when a person breathes on a window-pane in cold weather.

This moist condition of carpets and other fabrics is extremely favorable to the growth of microscopic fungi, and consequently equally injurious to human health. There is no cause for surprise that diseases of the respiratory system are so common.

The most important sanitary laws, respecting both dwellings and the human system, should be taught in our schools, as those who are now children will soon have the care both of their own lives and of the lives of the next generation.

If a spittoon containing the sputum of a person suffering from bronchial affections is put into a dark closet for twenty-four hours or more, it will be found, on examination, to have produced a growth of mould.

Put some mucus into a small vial and cork it loosely with a roll of paper; in two or three days a harvest of blue mould will be found in the vial.

Salt is death to microbes and fungi. If, then, salt or something equally destructive to parasitic animal and vegetable matter could without injury be taken into the lungs, in cases of consumption and other complaints of the respiratory system, it seems reasonable to suppose that it would prove beneficial.

Just as plants grow the entire year in conservatories, so our dwellings and our systems are hot-houses for the growth of mould.

Our exhalations are not composed entirely of air and vapor, but also of refuse organic matter which is thrown off through the lungs and, after drying, settles at last as a sediment on the floor, on articles of furniture or on our clothes.

Drain pipes often become clogged with the sediment from muddy water and refuse which passes through them, so that a force-pipe is required to remove the obstruction. In like manner our air passages frequently become stopped, and nothing but exertion or even violent exercise on our part will clear them.

Some may doubt the presence of unwholesome germs in the air, because they can not see them. It would be as reasonable to doubt the existence of the air itself simply because we can not see it.

Phlegm, or thickened mucus of catarrhal discharge, appears in some respects like freshly-mixed mortar. Water, lime, sand and hair are the components of mortar. In phlegm our system furnishes the water and a viscid, ropy substance, which takes the place of lime in mortar, while dust and fragments of mould and lint represent the sand and hair. The act of breathing helps to mix the ingredients, until the whole is of a uniform consistency. If the air were perfectly pure, no such thickened phlegm would be possible.

A fine gauze veil, worn over the mouth and nose,

is a safeguard against contagion, acting as a filter and catching most of the disease germs which would otherwise be inhaled.

Whatever part of the system air will reach, to that part also will poisonous germs, whether animal or vegetable, be carried.

A pint of water and half a pint of salt form a mixture that will kill grass or other vegetation on which it is thrown.

The tests and experiments given above are only a few of the many that might be cited in proof of what has gone before. They alone are given, however, because of their simplicity and ease of performance in any household.

www.ingramcontent.com/pod-product-compliance
Lightning Source LLC
Chambersburg PA
CBHW022006190326
41519CB00010B/1399